Sponsored Technology Workshops 2022

Held at the 2022 AIChE Annual Meeting

Phoenix, Arizona, USA
13-18 November 2022

ISBN: 978-1-7138-7921-3

Printed from e-media with permission by:

Curran Associates, Inc.
57 Morehouse Lane
Red Hook, NY 12571

AIChE
120 Wall Street, FL 23
New York, NY 10005-4020

Phone: (800) 242-4363
Fax: (203) 775-5177

www.aiche.org

Additional copies of this publication are available from:

Curran Associates, Inc.
57 Morehouse Lane
Red Hook, NY 12571 USA
Phone: 845-758-0400
Fax: 845-758-2633
Email: curran@proceedings.com
Web: www.proceedings.com

TABLE OF CONTENTS

Combining Simulation and Machine Learning for the Optimization of Powder Handling Processes with Altair Portfolio of Tools

Eric Veikle [a] Stefan Pantaleev [b]

[a] *Altair Engineering Inc., 1820 E. Big Beaver Rd., Troy Mi 4808, USA*
[b] *Altair Engineering Inc., 1 Rutland Court, EH3 8FL, Edinburgh, UK*

The trend in the pharmaceutical industry is to migrate towards continuous manufacturing, to achieve this a real-time model must be deployed to control the operational parameters of the process, to ensure product quality and maximize productivity. A common operation in the pharmaceutical industry is blending. In this study the development of a real-time fit-for-purpose model that was trained from high-fidelity DEM simulations will be discussed.

In the pharmaceutical industry manufacturers are presented with the challenge of ensuring the active ingredients in the medication is uniformly mixed prior to the formation of the delivery doses. Quantifying the uniformity of a mixture in a batch blender is a challenge, as it is experimentally opaque, and requires the interruption of the mixing process. Further, it is a challenge to obtain a representative sample that accurately depicts the degree of mixture in the entire bin blender. Instead of relying on empirical based models to predict the mixture uniformity it is possible to predict the mixture uniformity with DEM simulations [1]. The challenge with DEM simulations is their computational efficiency is quite low, simulations may take a day to complete. Therefore, DEM simulations on their own cannot provide the real-time model to control the mixing process. However, the DEM simulations can provide valuable insight into the evolution of the mixture as a function of operational parameters and that information can be used to train a neural-network model that has the required accuracy and ability to predict mixture uniformity in real-time.

With DEM simulations the fundamental challenge with modelling fine particulate solids such as powders at the particle level is the immense number of particles involved. One cubic centimetre of these materials could contain one trillion particles, which is too much to simulate in practice. A meso-scopic modelling approach can be used to address this challenge. The aim with this approach is to model the complex macro-mechanical behaviour of fine particulate solids using the micro-mechanical parameters but with virtual particle at an intermediate scale between the physical particle scale and scale of the system of interest. Therefore, a trade-off is made with this approach, whereby micro-mechanical fidelity is sacrificed in the interest of computational efficiency, but the bulk behaviour is modelled accurately. An additional benefit is the need to replicate the complex physical particle morphology is eliminated, because the meso-scopic particle represents an agglomerate of billions of physical particles that could take an arbitrary shape. This is a very useful simplification since characterising the physical particle morphology and replicating it in DEM models is extremely challenging and computationally expensive. Typically, the meso-scopic particle is modelled as a paired-sphere with a small aspect ratio instead. This simple and computationally efficient shape is found to result in sufficient shear dilation, which is an important phenomenon for fine bulk solids. It also eliminates non-physical packing crystallisation and has historically produced good results in different systems [2].

The meso-scopic approach has many useful features but there are also some challenges. It necessitates a sophisticated contact model to produce accurate results on the macro-scale. In this respect, EDEM utilises the Edinburgh-Elastic-Plastic-Adhesive contact model. DEM contact models are defined by micro-mechanical parameters which are difficult or impossible to measure. This is particularly true in the case of the meso-scopic modelling approach because the numerical particles represent physical particle agglomerates, and their material properties are not known a priori. For example, appropriate values for the particle shear modulus and Poisson's ratio need to be determined, along with interaction parameters such as friction coefficients and surface energy. Since these values cannot be directly measured, they must be calibrated.

In the context of DEM, calibration is the indirect determination of the micro-mechanical model parameters from a macro-mechanical response. A typical example is the static angle of repose test in which the angle that a conical pile of material forms with the horizontal is measured. This is a very common test for DEM model calibration due to its simplicity. The test procedure can be replicated in a DEM model and the model parameters can be optimized to reproduce the measured static angle of repose. This calibration process is the only practical way to produce a DEM model with an accurate bulk behavior and is a fundamental component of the DEM modelling methodology. However, calibration can be time consuming, and resource intensive due to the parameter variation. In certain cases, it can also be a complex, expertise intensive process. Furthermore, calibration is currently performed on a case-by-case basis with very little data exchange between DEM users. In this context a generic material model database has obvious utility it can minimize the effort and resource requirements and bring DEM model calibration within the reach of non-experts.

Therefore, the Powders Database was developed, which was created to address the challenge of quantifying the input parameters of the material model when using the meso-scopic modelling approach. In this case non-expert EDEM users can quickly quantify the input parameters by leveraging the 4,500 datapoint database and scaling laws which in theory give an infinite number of bulk densities and particle sizes within a wide range. The user must provide the poured bulk density, degree of compressibility, bulk cohesion along with either a static angle of repose measurement and or the shear stress response at two levels of normal stress. With this information the Powder Database will provide the best combination of input parameters for the material model. Though not all materials are represented in the Powder Database to sufficient accuracy, in these situations the user must still perform calibrations. The calibration of the input parameters of the material model used to represent the material in EDEM can largely be automated with the use of the EDEM-HyperStudy Connector. HyperStudy is Altair's design exploration tool that is used to create Design of Experiment (DoE), model fitting and optimization. The prepopulated EDEM decks included in the EDEM Calibration Kit include all the commonly adopted material characterization tests and python scripts to automate the extraction of material response under study. The availability of the Powder Database, the EDEM Calibration Kit and the ability to automate the calibration process with the EDEM-HyperStudy Connector reduce the time and cost associated with developing an acceptably accurate virtual representation of the physical material of interest.

High fidelity simulations were run in EDEM to predict the segregation index of the two-part mixture. The independent parameters that were included in the investigation were the mass of powder in the blender, the rotational velocity of the blender and the angle of twist, the orientation of the blender with respect to the axis about which the blender rotates. To minimize the number of high-fidelity slow solving EDEM simulations Altair HyperStudy was used to create an experimental design that would allow for quantification of first order effects between mass, rotational speed, and angle of twist on the segregation index. The Taguchi (3x3) experiment had the following design: the powder in the mixture was a two-part mixture of equal mass with levels of 1.6, 2.0 and 2.4 tonne. The rotational velocity of the blender was set at 8, 12 and 16 rpm. The angle of twist was set at 0, 22.5 and 45.0 degrees. In total 9 simulations were run in EDEM that were used to train the neural-network reduced order model.

The results of the simulations revealed that the initial mass of the powder had the greatest influence on segregation index, a direct relationship, followed by the rotation velocity, an indirect relationship and interestingly the angle of twist had very little effect on the segregation index. The direct relationship between initial mass and the segregation index was logical, the mixture would not be as uniformly mixed with an increase in initial mass of powder. The indirect relationship between rotational speed and segregation index was also logical, the mixture uniformity would improve with an increase in rotational velocity.

Altair's romAI was used to create and train the neural network model. The neural network was trained with the nine simulations, including the independent parameters and the evolution of the segregation index with time. Prior to training the segregation index response from EDEM was filtered to remove high frequency variation with a low pass filter set at 0.1 Hz as noise poses challenges for neural network training. A non-linear architecture with 2 hidden layers and 160 neurons was trained. The training parameters used were 1000 Epochs, with a test split ratio of 0.01 and a cross-validation split ratio of 0.01. The accuracy of the trained model can be assessed by comparing the predicted segregation index to an unfamiliar independent dataset. Figure 1 shows the comparison of the segregation index predicted from the trained romAI model and EDEM with the following bin operational parameters: mass of 2.24 tonne, 14.7 rpm and 34.4 degrees. The agreement between the romAI trained model and the high fidelity EDEM simulation are quite good.

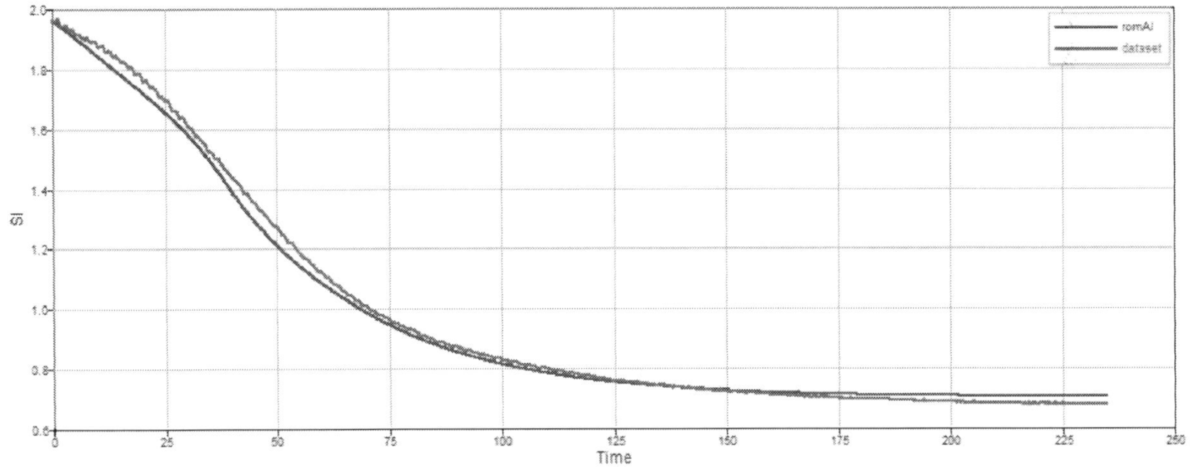

Figure 1 Evolution of segregation index as a function of time in seconds from the romAI trained model and EDEM using operational parameters that were within the design space of the DoE but not used to train the romAI model.

The romAI model of the bin blender was then imported into Activate, Altair's system modeling software. The operational parameters of the bin blender were optimized using the gradient-based optimizer which operates on the normalized parameters for statistical purposes. The normalization of all the parameters is automated in Activate. This approach allows for a large number of parameters to be investigated in real time; the segregation index was instantaneously predicted for each set of operational parameters using the romAI model. In this case 50 iterations were run in a few seconds and an optimum was achieved. The optimum parameters were a mass of 1.86 tonne, rotational speed of 9.8 rpm and an angle of twist of 19.0 degrees. The prediction of the segregation index evolution from the romAI model was validated against the high fidelity EDEM simulation using the same optimum operational parameters as shown in Figure 2. Again, there is very good agreement between the two curves.

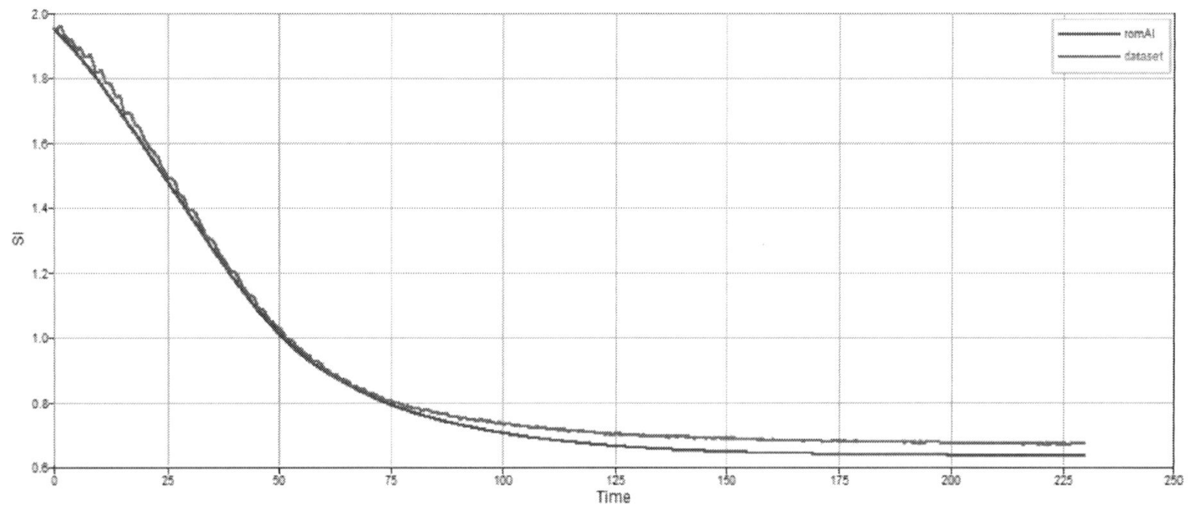

Figure 2 Evolution of segregation index as a function of time in seconds from the romAI trained model and EDEM for the optimum operational parameters of the bin blender.

This research study demonstrated the utility of the combination of high-fidelity DEM simulations with strategic experimental design, which was used to train a neural network reduced order model allowing for the optimization of the operational parameters of a bin blender.

Keywords: Bin Blending, Optimization, Reduce Order Model, Real-Time Simulation, DEM, Machine Learning

References

[1] Lemieux, M., Léonard, G., Doucet, J., Leclaire, L.-A., Viens, F., Chaouki, J., Bertrand F. (2008). Large-scale numerical investigation of solids mixing in a V-blender using the discrete element method. *Powder Technology, 181, 205-216.*

[2] Pantaleev, S., Yordanova, S., Janda, A., Marigo, M., & Ooi, J. Y. (2017). An experimentally validated DEM study of powder mixing in a paddle blade mixer. *Powder Technology, 311*, 287–302.

716a Beyond the Optimizer: Gurobi's Hidden Gem Features and Applications in the Chemical Industry

Ed Rothberg

Optimization is a foundational tool for many aspects of process design andamp; operation, from predictive real-time control schemes to long-term planning models. The ability to formulate a mathematical model of the problem at hand is what unifies these applications across a range of time scales and industries. The business or engineering problem may be yours, but the science of decision intelligence is Gurobi's – let us help you solve it.

This workshop will dive into powerful features that would be of interest to anyone using or creating optimization applications, or those who want to learn more about turning rules-of-thumb into solutions with guaranteed global optimality. We provide tools to make it easy to incorporate multiple objectives, scenario analysis, nonlinear functions, and more into your optimization models. These features can streamline your modeling efforts and inspire new application ideas. There are also features like warm starts, variable partitioning, and more that can enhance the solver performance. We will discuss use cases for how these may be applied, from upstream field planning, to downstream blending, to environmental studies.

We'll also review a number of other important ingredients in a successful optimization application, including support, licensing, education, consulting, etc. This will include a short demo of the Burrito Optimization Game (BurritoOptimizationGame.com), the educational resource that introduces players to the power and capabilities of mathematical optimization.

719a Chemical recycling, a big step toward a smaller footprint

Jayme Leita

Challenges for innovation have become more complex as companies are striving to meet the increasing demands for performance while addressing sustainability macrotrends. During this interactive session you will hear how Eastman, a materials innovation company, is developing more sustainable materials while influencing the broader eco-system to ensure their long-term adoption. Attend for a chance to win prizes made from our Tritan Renew Copolyester

718a Combined In-Silico and Experimental Approach for Solid Form Crystallization Development

Yuriy Abramov, Shanming Kuang

This presentation will focus on combined in-silico andamp; experimental approaches for solid form screening andamp; selection, including validated methodology and workflow. To allow for a rational approach to API crystallization development, in-silico screening approaches are discussed. Such approaches include virtual solvent screening for drug substance crystallization, solid state de-solvation, deagglomeration/agglomeration, shape optimization and impurities rejection. Application of rational co-former or counterion screening will be discussed as well to guide solid form selection/switching for improved physical properties of drug substance. Examples of combined in-silico andamp; experimental selections of solvent and co-former or counterion for crystallization development will be presented.

727a Developing Industry Ready Skillsets for Engineers

Walter Coto

In this session, join Elsevier's Engineering Consultants as they discuss their transition to the Engineering Industry, highlight some of the skills and expectations companies across different industries have of students, and discuss how engineering skills can be applied to other areas to grow your career. Learn ways to build your non-engineering skills to make the best impression at interviews, and ways to continue to grow your skill set as a working engineer once you've found a company to grow with.

728a Scale down first, before you scale up - a take on scale up of industrial microbial fermentation process

Sangram Yadav

Scale-up of the microbial fermentation process is one of the most critical aspects for a company to commercialize its products.

Leverage the know-how of your own manufacturing plant or a tolling facility to create a better scale-down model. This will improve the probability of successful scale-up and the ideal time in the process development phase to initiate scale-up-related discussions.

729a Computational Chemistry Workloads via HPC Built for the Cloud

Clinton Smith, Khatereh Kashmari

Computational simulation continues to play an increasingly critical role in the understanding and development of products and processes within the life sciences and energy sectors.

Advancements in high performance computing (HPC) in the cloud have unlocked acceleration opportunities for computational fluid dynamics, computational chemistry and molecular dynamic simulations to discover new drugs and materials.

Rescale provides access to over 100+ HPC architectures in a multi cloud environment. These include NVIDIA GPU Architectures to further accelerate workloads. This plethora of HPC infrastructure delivers the promise of right-sizing the software workload with the most cost-performant hardware, enabling computational chemistry, molecular dynamics and computational fluid dynamic (CFD) workloads from open source, custom/proprietary and commercial software vendors.

Workflows enabled by software toolkits such as Python, R, MATLAB and others are also supported within Rescale's multi cloud environment.

In this workshop, we'll provide a demo walkthrough of running computational chemistry codes such as GROMACS, LAMMPS or Quantum Espresso on the Rescale Platform.

"Through Rescale, we were able to draw results more than 2x faster than our existing workflow, which helped us make better business decisions."
- Jerry Maeng Managing Director, AZothBio

"Bionano's use of Azure on Rescale reduces compute cost for genetic disease analysis by more than 50%, compute time by 30%."
- Mark Oldakowski, COO, Bionano

Simulate pharmaceutical processes and plants to accelerate development

Ravindra Aglave, Ananya Chowdhury

Discover all the tools you need to help pharmaceutical companies bring high quality drugs to patients faster by developing, applying and deploying science-based modeling technologies. In recent years, digital twins have been applied to a broad range of applications in pharmaceutical Randamp;D and engineering, improving efficiency through reducing experimental requirements, facilitating scale-up and tech transfer, and enabling virtual design space exploration. Our comprehensive approach to digital twins integrates simulation tools that help pharmaceutical companies optimize the entire product lifecycle, from research to production.

Join this workshop and learn how:

- Computational methods at molecular and mesoscales can help develop and screen thousands of new chemical formulations for new materials, solvents and bio pharmaceuticals through property estimations
- Digital twins for continuous manufacturing of synthetic drug substance can be used for the development and testing of advanced process control strategies
- CFD (Computational Fluid Dynamics) can be employed for making well-informed decisions in scaling-
up and optimization of multiphase reactors
- Shelf lives of pharmaceutical ingredients can be elongated using CFD indoor environment modeling
- End-to-end bioprocess digital twins can be used for faster and more efficient process design and optimization, risk assessment and scale up
- System models for pharmaceutical manufacturing processes can be deployed across the organization for accelerating development to launch lead times.

AUTHOR INDEX